What Is Astronomy?

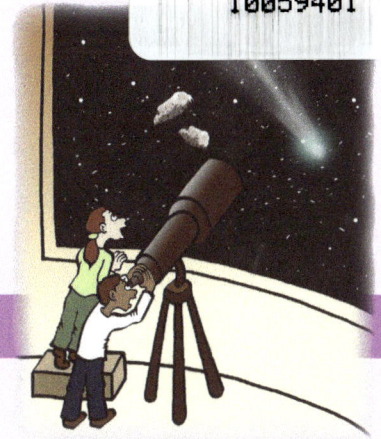

Illustrations: Janet Moneymaker
Design/Editing: Marjie Bassler

What Is Astronomy?
ISBN 978-1-950415-38-0

Published by Gravitas Publications Inc.
Imprint: Real Science-4-Kids
www.gravitaspublications.com
www.realscience4kids.com

Do you ever go outside and look up at the night sky?

Yes! I do!

If it is a clear night and you don't live in a big city, you can likely see many **stars**.

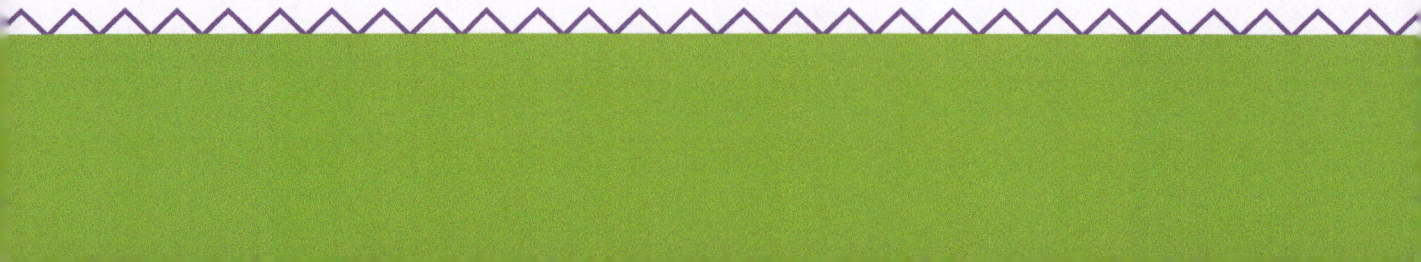

Astronomy is the study of stars and all the other objects that are far away from Earth in outer space. **Astronomers** are the **scientists** who observe objects in space.

I want to be an astronomer!

It is hard to say who were the first astronomers. Many early people in ancient times studied the sky and the movement of stars, the Moon, and the Sun.

I see the Moon coming up in different places.

Many early people were able to track the Sun and Moon to create calendars.

I wonder how that works.

An early Mayan calendar

Early astronomers also gave names
to individual stars and groups of stars.

That one looks like a mouse!

This one looks like cheese!

Astronomers would like to be able to fly into space to visit stars and other faraway objects. Instead, they use tools to study the objects.

The **telescope** is one type of tool used by astronomers to look deep into outer space.

A **telescope** is a tool that is used to make faraway objects look bigger, or **magnify**, them. One type of telescope uses **lenses** to magnify objects in space.

Telescope

The lens you look through

The lens that points toward the object

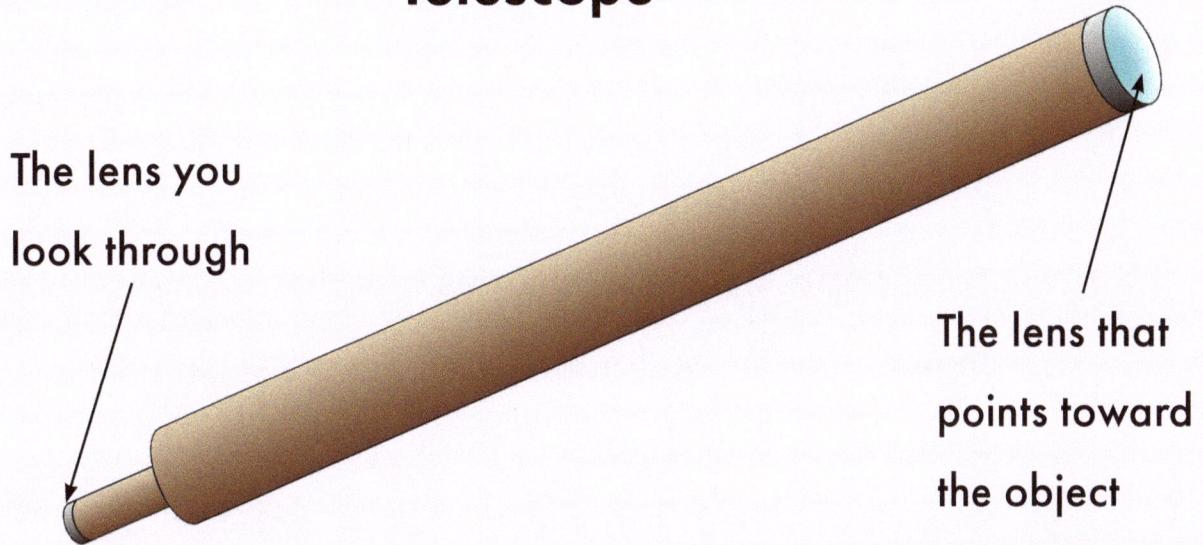

A **lens** is a piece of clear glass or plastic that is shaped in a way that causes it to magnify objects.

Another type of telescope is a **space telescope**. Space telescopes are placed high above the surface of Earth. They travel around Earth and gather information about objects in space. This information is sent back to astronomers on Earth.

Hubble Space Telescope

Astronomy is a very exciting and rapidly changing field of scientific study. New discoveries are constantly being made. And there is so much more to be learned and explored.

Wow! Look at all that stuff out there!

How to say science words

astronomer   (uh-STRAH-nuh-mer)

astronomy   (uh-STRAH-nuh-mee)

calendar   (KAA-luhn-duhr)

Hubble   (HUH-buhl)

lens   (LENZ)

magnify   (MAG-nuh-fiy)

Mayan   (MIY-uhn)

science   (SIY-uhns)

scientist   (SIY-uhn-tist)

space   (SPAYSS)

telescope   (TE-luh-skohp)

What questions do you have about ASTRONOMY?

Learn More Real Science!

Complete science curricula
from Real Science-4-Kids

Focus On Series

Unit study for elementary and middle school levels

Chemistry
Biology
Physics
Geology
Astronomy

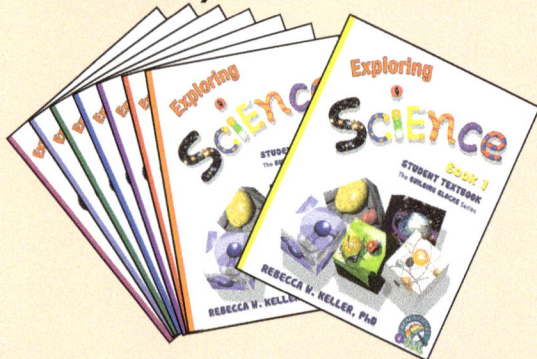

Exploring Science Series

Graded series for levels K–8. Each book contains 4 chapters of:

Chemistry
Biology
Physics
Geology
Astronomy

www.ingramcontent.com/pod-product-compliance
Lightning Source LLC
Chambersburg PA
CBHW040153200326
41520CB00028B/7591